FRUITS

A Crabtree Roots Book

ALICIA RODRIGUEZ

Crabtree Publishing
crabtreebooks.com

School-to-Home Support for Caregivers and Teachers

This book helps children grow by letting them practice reading. Here are a few guiding questions to help the reader with building his or her comprehension skills. Possible answers appear here in red.

Before Reading:

- What do I think this book is about?
    - *This book is about fruit.*
    - *This book is about what fruits look like.*
- What do I want to learn about this topic?
    - *I want to learn where fruits grow.*
    - *I want to learn how big fruits can be.*

During Reading:

- I wonder why...
    - *I wonder why plants make fruit.*
    - *I wonder why people eat fruit.*
- What have I learned so far?
    - *I have learned that fruit can be big or little.*
    - *I have learned that fruit can grow in trees.*

After Reading:

- What details did I learn about this topic?
    - *I have learned that fruit can grow on vines.*
    - *I have learned that not all plants make fruit.*
- Read the book again and look for the vocabulary words.
    - *I see the word **fruits** on page 3 and the word **vines** on page 8. The other vocabulary word is found on page 14.*

These are **fruits**.

Some fruits are big.

Some fruits are little.

Fruit can be on **vines**.

Fruit can be in **trees**.

Most fruit you can eat!

Word List

Sight Words

are	little	these
be	most	you
can	on	
eat	some	

Words to Know

fruits

trees

vines

26 Words

These are **fruits**.

Some fruits are big.

Some fruits are little.

Fruit can be on **vines**.

Fruit can be in **trees**.

Most fruit you can eat!

PARTS OF A PLANT

FRUITS

Written by: Alicia Rodriguez
Designed by: Rhea Wallace
Series Development: James Earley
Proofreader: Ellen Rodger
Educational Consultant: Marie Lemke M.Ed.

Photographs:
Shutterstock: Nitr: cover; Mazur Travel: p. 1; Armas Vladimir: p. 3, 14; Alaa AbuMadi: p. 5; Mariola Anna S: p. 6; Lili Graphie: p. 9, 14; images72: p. 10-11; Samuel Borges Photography: p. 13

Crabtree Publishing

crabtreebooks.com 800-387-7650

In Canada: We acknowledge the financial support of the Government of Canada through the Canada Book Fund for our publishing activities.

Printed in Canada/062024/CP20240617

Published in Canada
Crabtree Publishing
616 Welland Avenue
St. Catharines, Ontario
L2M 5V6

Published in the United States
Crabtree Publishing
347 Fifth Avenue
Suite 1402-145
New York, NY 10016

Hardcover	978-1-4271-4065-4
Paperback	978-1-4271-4071-5
Ebook (pdf)	978-1-4271-3334-2
Epub	978-1-4271-3394-6
Read-along	978-1-4271-4077-7
Audio book	978-1-4271-3911-5

Library and Archives Canada Cataloguing in Publication

Title: Fruits / Alicia Rodriguez.
Names: Rodriguez, Alicia (Children's author), author.
Description: Series statement: Parts of a plant | "A Crabtree roots book".
Identifiers: Canadiana (print) 20210176156 | Canadiana (ebook) 20210176172 | ISBN 9781427140654 (hardcover) | ISBN 9781427140715 (softcover) | ISBN 9781427133342 (HTML) | ISBN 9781427133946 (EPUB) | ISBN 9781427140777 (read-along ebook)
Subjects: LCSH: Fruit—Juvenile literature.
Classification: LCC QK660 .R63 2022 | DDC j575.6/7—dc23

Library of Congress Cataloging-in-Publication Data

Names: Rodríguez, Alicia, 1992- author.
Names: Rodriguez, Alicia (Children's author), author. | Rodriguez, Alicia (Children's author) Parts of a plant (Crabtree Publishing Company)
Title: Fruits / Alicia Rodriguez.
Description: New York, N.Y. : Crabtree Publishing, 2022. | Series: Parts of a plant | "A Crabtree roots book.". | Includes index.
Identifiers: LCCN 2021014300 (print) | LCCN 2021014301 (ebook) | ISBN 9781427140654 (hardcover) | ISBN 9781427140715 (paperback) ISBN 9781427133342 (ebook) | ISBN 9781427133946 (epub) | ISBN 9781427140777
Subjects: LCSH: Fruit--Juvenile literature. | Picture books for children--Juvenile literature.
Classification: LCC SB357.2 (ebook) | LCC SB357.2 .R64 2022 (print) | DDC 581.4/64--dc23
LC record available at https://lccn.loc.gov/2021014300